YOUR KNOWLEDGE HAS VALUE

- We will publish your bachelor's and
  master's thesis, essays and papers

- Your own eBook and book -
  sold worldwide in all relevant shops

- Earn money with each sale

Upload your text at www.GRIN.com
and publish for free

Bibliographic information published by the German National Library:

The German National Library lists this publication in the National Bibliography; detailed bibliographic data are available on the Internet at http://dnb.dnb.de .

Imprint:

Copyright © 2016 GRIN Verlag, Open Publishing GmbH
Print and binding: Books on Demand GmbH, Norderstedt Germany
ISBN: 9783668336995

This book at GRIN:

http://www.grin.com/en/e-book/343280/gas-tentacles-russian-domination-of-european-gas-markets

Harel Tanjong

Gas Tentacles. Russian Domination of European Gas Markets

GRIN Publishing

Gas Tentacles:

Russia's domination of European gas markets

And Solutions

Abstract

This report examines the geopolitical situation between Europe and the Russian Federation. Observers have noted the tepidness of some European nations to oppose Russian interest. We believe this is due to Russia's comparative advantage in the energy sector, specifically Natural Gas. This resource, abundant in Russia, is the cornerstone of European energy use especially for nations of the former Soviet Union, creating economic pressure for them to agree to Russian demands. Being an abundant and relatively clean resource, natural gas is a preferred source of energy. Control of Russian supplies is dominated by state-owned Gazprom, which is the largest company in the country and exerts significant influence in the Russian Government. This report will also examine solutions for Europe's energy independence and security through strategies such as anti-trust action and new investments in domestic production.

Key Words: **Russia, Geopolitics, Natural Gas, European Energy**

Content

<u>**Thesis:**</u>

Being one of the most widely used and clean fuels on the planet, natural gas is a key 21st century resource. Used for everything from cooking to heating, some countries cannot go without it. Fortunately for the Russian Federation, it can be found in abundance on its lands. The administration of Vladimir Putin has exploited it rigorously for the financial benefit of Russia, an understandable act of self-interest. Who is Russia's biggest customer? Nearby continental Europe. **Remarkably, this earth based resource has given Russia a political and financial stranglehold on its European neighbors. It is time for Europe, for its own sake, to diversify and invest in its supply of natural gas.**

<u>**Importance of Natural Gas:**</u>

For centuries, natural gas has been utilized all over the globe. It can be found on almost every continent, and is a very versatile energy source. Natural Gas is the cleanest burning alternative fuel. Automobiles today that utilize it emit less pollutants than the standard petroleum car (Demibras 2006). With the acknowledgement of climate change by global scientific and political bodies, there have been increased calls for the use of Natural Gas. Being predominantly made up of methane, it emits less CO2 into the atmosphere than coal or oil. Combined with the benefits of being clean, NG is an abundant fuel. Found worldwide, its numbers have been on the increase since the 1970s.

In its early years natural gas was mainly used to power street lamps. Today, it is one of the most widely used fuels. It has been utilized for heating and cooling homes and businesses, as well as water heating. NG makes for a very good cooking option. For the average consumer it is one of the cheapest fuels available. Additionally, "Cooking with an NG range or oven can provide many benefits, including easy temperature control, self-ignition and cleaning, as well as being one half the cost of cooking with an electric range" (Demibras 2006, 419). The gas is also being heavily utilized for electricity. The generation of electricity from the gas typically comes from a boiler-to-steam process. Natural Gas electricity plants are one of the cheapest power plants to build, and the industry is increasingly turning to the fuel as a result (Union of Concerned Scientist 2016). All of these factors have contributed in forming a wide and growing market for Natural Gas.

<u>**Natural Gas In Russia:**</u>

Gas development in Russia started at the end of the 19th century, with help from a few wealthy investors such as the Rothschild family (Victor 2008). Gas networks were eventually

3

established in all the main cities, and were mainly utilized for home lighting. Under the leadership of Nikita Khrushchev, who led the Soviet Union from 1953 to 1964, the country accelerated its gas production (Victor 2008). This was part of Khrushchev's ambitious goal to catch the United States economically within 25 years (Victor 2008). As gas reserves west of the Urals became depleted (because of their proximity to the main markets), attention turned to developing the fields in Siberia. This came to be known as the "Siberian Period" (Victor 2008).

Today, natural gas can be found all over the world, but the largest reserves are found in the former Soviet Union, particularly the Russian Federation (EIA 2015). With 47,573 reserves, it doubles the amount of the next closest nation, Iran (Schenk 2012). Most of its reserves and centers of production are found in northern West Siberia. However, other locations are being scouted for NG including the Yamal Peninsula, Eastern Siberia, and Sakhalin Island (EIA 2015). According to the U.S. Geological survey, Russia has the most untapped natural gas reserves of any country (Schenk 2012). There are also believed to be a healthy amount of gas reserves in the arctic, which, for the moment, remain too cost intensive to extract due to the climatic harshness of the region (Schenk 2012). However, climate change, and the rapid melting of sea ice, make this a lucrative venture for the future.

Gazprom:

Oil and Gas were the financial saviors of Russia beginning in the 90's after the breakup of the Soviet Union. With a financially impaired federal government, Russia looked for a way to raise funds and reduce its deficit. Oil was completely privatized, providing a solid tax base, and appeasing numerous Kremlin affiliated oligarchs (Victor 2008). The government took a different approach with Natural Gas, however. It kept its significant stake in Gazprom, the unquestioned head of the natural gas industry in the country. A corporate behemoth, it currently controls 75 % of the country's reserves and 94 % of its production (Victor 2008). Gazprom today is considered, by market capitalization, the richest company in Russia. The firm's extensive wealth has allowed it to enjoy significant political support and influence (Victor 2008).

Gazprom has had a long history of relations with the Russian/Soviet state. It was founded as the Soviet Gas Ministry by the Communist Party. With the breakup of the Soviet Union and subsequent privatization of the industry in the early 90's, it became a joint stock company (Victor 2008). The company quickly developed a monopoly on the natural gas industry. With a significant percentage of its shares in government hands, anti-trust action did not happen.

Government favoritism for the company also stemmed from the fact that the oil industry was mostly in private hands, with much of the profits in that sector were being moved offshore. The Russian government thus saw gas as a financial imperative (Victor 2008).

By 2000, the industry still wasn't as controlled as the government would have liked. The rise of global energy prices triggered an urgency of the new Putin administration to establish stronger control, specifically over Gazprom. Old oligarchs and Soviet-era bureaucrats were removed, in place of more favorable Kremlin allies (Heinrich 2008). The government also increased its share of the company to 51 % in 2005 (Heinrich 2008). The Russian government, however, was not interested in having its precious company exposed to the market. The one regulation standing in the way of complete Gazprom control was the anti-monopoly services regulation, which was eventually repealed (Victor 2008). Gazprom was then permitted to buy up any domestic gas producing enterprise it wanted, without fear of legal reprimand (Victor 2008).

The company enjoyed more privileges for being so closely associated with the Russian government. By supplying low price gas to domestic consumers, the Russian economy as a whole, benefitted. In return for this, the Government gave Gazprom exclusive access to hard currencies from its exports. This allowed the company to finance much of its own investments, becoming almost a "state within a state" (Victor 2008). A "revolving door" of sorts has developed between the new institutions, with many members of the Gazprom board being former high ranking officials in the Russian Government (Baran 2007). Nowadays, policy of the Russian Government and Gazprom are almost in lockstep. The gas giant is used as a tool by the Kremlin to achieve its foreign policy objectives. Putin is now in de facto control of the company, while the CEO is Alexey Miller, a former Putin aide. In fact, it was the Russian president, not the Gazprom CEO, who announced that there would be no more gas discounts for Ukraine (Bryce 2014).

Russia's Geopolitical Influence:

In Europe, a cohesive energy policy/ strategy is virtually nonexistent. Matters of energy security are often left to individual states or the free market (Baran 2007). Russia has taken advantage of this lack of cohesion to secure top deals for its gas exports. Many European countries, particularly the former Soviet republics, are heavily reliant on gas for energy. This is especially the case during the very cold eastern European winters (Gawdhat 2006). European attempts to construct routes that circumvent Russia by passing through central Asia have largely been blocked by the Russians (Baran 2007). Russia imports its gas from its former Soviet Central Asian neighbors at below market prices. Much of that same gas is then exported to Europe, where citizens there pay the full market price (Schaffer 2008). Gas sold domestically in Russia is heavily subsidized, so the Europeans are made to make up for the difference (Schaffer 2008).

The most important pipelines utilized by Gazprom run from Siberia into Europe through Ukraine (Schaffer 2008). There is also the Yamal-Europe pipeline which runs through Belarus, (The Economist 2007) then into Poland and Germany. In the works is the Nord Stream Pipeline, which will pass through the Baltics and connect with the Yamal-Europe Pipeline in Germany, avoiding transit fees from Belarus along the way (Schaffer 2008). This pipeline is expected to cost at least $10 billion (Baran 135), which would be a large sum for most other energy companies, but not one that is majority state owned, and shielded form most market forces.

Vladimir Putin has successfully leveraged Russia's economy of scale with gas as a political tool to influence Europe. Presently, there is no ready alternative to Russian supplies. The Europeans are hooked on what the supplier has to offer, "Thus if a supplier refuses to supply gas or charges an unreasonable price, the consumer cannot quickly or easily turn to another source. The consumer state would have no choice but to accept the suppliers' conditions or go without natural gas, an option that is all but unacceptable for most" (Baran 2007). The European Union overall relies on Russia for half of its natural gas imports, with percentages higher in Eastern Europe (Baran 2007). Latvia and Lithuania are prime examples of what can happen when Russia decides to cut off energy supplies. In 2003, Russia stopped supplying oil to Latvia after its government refused to sell one of its export facilities to a Russian company. Lithuania was punished after a Russian firm operating in the country was denied access to critical infrastructure (Baran 2007). Russia often cuts supplies while blaming the disruption on "technical difficulties." Despite this, third parties are often refused to come to the sites to examine the issues. The "difficulties" typically go unresolved until the host country/firm agrees to Russian demands (Baran 2007).

Gazprom (and simultaneously, the Russian Government) has sought to deepen Europe's dependence on it by buying significant shares of European firms. Knowing signing a deal with Gazprom will bring access to greater markets and profits, the European firms have been played against each other by Russia (Baran 2007). The Russians have essentially used Europe's strength, the free market, against it, for great benefit. To date, Gazprom has made deals with some of Europe's largest energy companies, including; E.ON (Germany), DONG Energy (Denmark), Eni (Italy), and BASF (Germany) (Victor 2008).

The foreign policy goals of the EU also suffer with this state of dependence, with Europe often turning a blind eye to many Russian misdeeds. The aforementioned supply cuts to Latvia and Lithuania drew little criticism or derision from the rest of the continent (Baran 2007). Western European nations did not want to risk losing their bargaining chip with Russia over the tiny Baltic States (Baran 2007). Potential access to huge Russian reserves of gas was too lucrative. These differences of interest are causing a split between east and west.

The European nation with possibly the most energy conflict with Russia in the 21st century has been Ukraine. The two countries have been involved in four major gas disputes since 2005 (BBC 2009). Compliance from Ukraine is so crucial to the Russian strategy because 80 percent of Russian exports to the European Union pass through the country (BBC 2009). In 2009, Gazprom reduced the supply of Russian gas bound to Ukraine due to outstanding debts from the Ukrainian government of 2.4 billion. All gas flows were halted for about two weeks, which also cut off supplies to southeastern Europe (Simon 2009, Stern 2009, Yafimava 2009). Later in 2014, after the Crimea crisis, the Crimean authorities attempted to nationalize their firm Chornomornaftogaz. The firm owed debts to Ukrainian companies, but the Crimeans considered the debts null and void, and decided Gazprom should take over the company (Interfax 2014). These moves were as much about politics as they were about money. Russia has always been keen to keep Ukraine in its sphere of influence. After the Orange revolution and Ukraine's intentions to join NATO, Russia was displeased, and felt it needed to send a message by cutting off a key component of Ukraine's economy (Simon 2009, Stern 2009, Yafimava 2009).

The European reluctance to challenge Russia for its anti-market actions has negative consequences for its citizens. Russian inefficiencies in extraction and delivery of natural gas make consuming it more expensive for Europeans (Schaffer 2008). With a virtual monopoly on NG in Russia, Gazprom has little incentive to improve or reform its practices (Baran 2007). "Between 1998 and 2005, output in Russia's then mostly privately owned oil sector rose by 50 percent. During that same period. Production in the gas sector (Gazprom) barely grew at all. Since 2004, when the Kremlin began its consolidation over the oil sector in earnest, Russian oil production has leveled off as well." (Baran 134) Lack of competition, and smaller supply are hurting European energy consumers. The citizens of the continent are literally paying the price for Gazprom's monopolization and inefficiencies.

Recently, because of sanctions placed on it by the Crimean crisis, Russia has turned to another large potential customer market in China. The new agreement with China has been seen as a major victory for Russia, allowing it to diversify its market, and reduce the effect of recent western sanctions (Perlez 2014).

Possible European Strategies for Energy Independence:

At the moment, the European nations seem like hopeless energy consumers to Russia, but there are possible strategies for these countries to move toward energy independence. Russia may be pursuing an assertive foreign policy in recent times, but this should not serve as a pretext for inflamed rhetoric or western troops to be sent to its borders. A militaristic response would only escalate tensions. It would be wiser of Europe to pursue a strong financial approach and offer states neighboring Russia an alternative. Europe must also not assume that Russian gas will be flowing to it forever. There have been concerns of shortages in the main extraction fields (Victor 2008). Maintaining production would require new large investments on the part of Gazprom in Siberia (Baran 2007). Europe should not rely on this, any major disruption of gas flows into the continent could produce negative economic shocks. Major slowdowns in industry and business, which rely heavily on gas power, could occur (Heinrich 2008). A sudden price rise from a supply drop could send many important companies out of business in these areas. Equally as bad, gas prices for the average European homeowner would also skyrocket, crashing discretionary spending and forcing some into desperate measures to stay warm during the winter months (Heinrich 2008).

The area around the Caspian Sea is critical for a strategy of European energy diversion from Russia. The area has a huge future potential for oil and gas (Schaffer 2008). Joint energy projects with nations such as Turkmenistan, Azerbaijan, and Georgia are possible and have been done before (Baran 2007). Also, the E.U. can utilize its relationship with the United States. The U.S. played a critical role in starting construction of a Caspian energy pipeline that circumvented Russia. (Baran 2007) Russian response was bellicose as predicted, but the project was eventually completed. The Baku-Tblisi-Ceyhan pipeline runs from Baku, in Azerbaijan, through Georgia, and ends up in Turkey, without once touching Russian territory or being subject to its transit fees (Baran 2007). The Nabucco is another pipeline project utilizing the collaboration of Europe and the U.S (Baran 2007). This will again tap into the vast potential gas resources of Azerbaijan, although Russia has taken action to disrupt the process, swaying countries on the routes proposed path by making deals with them. The pipelines that have been completed not only benefitted Europe, but the nations that they passed through. With new, positive interactions with western institutions, Georgia and Azerbaijan are poised to further gravitate toward the modern European community (Baran 2007).

European companies are underrepresented in the growing Central Asian gas market. It is vital that the E.U. continues to engage countries in the area in order to give companies such as Royal Dutch Shell or BP the confidence to invest there (Baran 2007). The C.E.Os of these companies will not risk throwing out billions of dollars on these projects if the E.U. is not willing to compete in the region (Baran 2007).

Forcing Russia and Gazprom towards a more market based approach to business will loosen their grip on Europe. The E.U. possesses the ability to prosecute large corporations such as Gazprom on anti-trust grounds (Baran 2007). It has already taken action on General Electric and Honeywell for what it saw as monopolistic dealings (Baran 2007). This action will also address the aforementioned issue of efficiency. Cost will be lowered for European consumers by making extraction and distribution across the continent more exposed to market pressure and incentive (Schaffer 2008).

Membership in the World Trade Organization is a possible source of leverage that the E.U. can use with Russia. Russia has previously expressed a desire to join the WTO (Schaffer 2008). Members of this organization, however, have to agree to more liberalized trade rules. The WTO dictates that all countries must conduct business with each other with lower tariffs, import quotas, and regulations (Schaffer 2008). Russia would also have to ratify the Energy Charter Treaty, signed by 53 different countries. If Russia were to ratify this legislation, it would no longer be able to monopolize gas routes entering Europe.

The disunity among European states with regards to its energy strategy put many of them in a state of dependence to Russia. If Europe is serious about being a singular political union, it must adopt a common energy strategy. All nations should balance their own energy priorities with those of the continent. This will take significant political commitment, commitment that some might be unwilling to make, but it must be done in order for the current situation to change. European states must realize that this common goal can benefit them all in the long term. The Schengen Agreement is a prime example of European political unity. 26 European states are signatories of this agreement, which eliminates border controls and establishes a common visa framework, allowing for free travel throughout the signatory nations (European Commission 2011) . This shows that meaningful collaboration on the continent is possible Russia has been able to achieve such a foothold because of its decisiveness and concrete proposals. E.U. energy policies need to be streamlined in order to compete with Russia in the gas game.

Despite moving towards a more independent stance, Russia cannot be entirely eliminated from the E.U.'s energy picture, nor should it. Its influence should simply be reduced to level playing fields with other European firms to give consumers adequate choice. The country possesses a vast amount of resources, much of them untapped. Gazprom, facing increasing demand at home and in Europe, and lacking investment in new extraction infrastructure, has been forced to import more quantities from Central Asia (Heinrich 2008). The E.U. with its larger financial capital and strength has the ability to develop untapped areas of Siberia and Central Asia for its own energy needs. E.U.-Russia dialogue has resulted in the creation of the Skoltech Center for Energy Systems in Moscow, along with proposals to improve the safety of transporting of oil and gas (Bahghat 2006). According to a recent study by the IEA, Russia's gas industry requires investment of up to 11 billion a year (Bahghat 2006), a figure that Russia may not be able to cover by itself. Cooperation is possible, but it will once again take assertiveness and decisiveness amongst European leaders to make sure that the Continent's gas needs are fulfilled.

Conclusion:

Russia's blessing with an abundance of Natural Gas has lifted the country out of its dire early post-soviet condition. With the admission of current President Putin himself, Russia has used this resource, in high demand, to reassert itself over its neighbors. Putin's Russia does not want to see itself left behind by its former Soviet republics as well as the international community at large. This has large implications for Europe, dependent as it is on Russian gas. This passive dependence is politically and economically detrimental, and the European community must take active steps to diversify its energy sources and intake. A common energy framework must be developed, uniting Europe with one voice on the matter. Europe must engage with energy rich nations in central Asia, using its financial clout to secure future energy investments. Finally it is essential for the European Union and its allies to push for more market based reforms in Russia. Using its legal levers with anti-trust laws, Europe can, and must, break Russia's monopoly grip on its gas supplies.

<u>References</u>

Bahgat, Gawdhat. "Europe's energy security: challenges and opportunities." *International Affairs* 82 (2006).

Baker Schaffer, Marvin. "The great gas pipeline game: monopolistic expansion of Russia's Gazprom into European markets." Foresight 10, no. 5 (2008): 11-23. doi:10.1108/14636680810918478.

Baran, Zeyno. "EU Energy Security: Time To End Russian Leverage." The Washington Quarterly, October 2007.

British Broadcasting Corporation. "BBC NEWS | Europe | EU Reaches Gas Deal with Ukraine." Home - BBC News. Last modified August 1, 2009. http://news.bbc.co.uk/2/hi/europe/8179461.stm.

Bryce, Robert. "National Review Online." National Review Online. Accessed March 18, 2016. http://www.nationalreview.com/article/372804/russia-gazprom-robert-bryce.

Demirbas, Ayhan. "The Importance of Natural Gas as a World Fuel." Energy Sources, December 2006.

Heinrich, Andreas. "Under the Kremlin's Thumb: Does Increased State Control in the Russian Gas Sector Endanger European Energy Security?" Europe-Asia Studies 60, no. 9 (2008): 1539-1574.

Perlez, Jane. "China and Russia Reach 30-Year Gas Deal - The New York Times." The New York Times - Breaking News, World News & Multimedia.2016. http://www.nytimes.com/2014/05/22/world/asia/china-russia-gas-deal.html?_r=0.

Pirani, Simon, Johnathan Stern, and Katja Yafimava. "The Russo-Ukrainian gas dispute of January 2009: a comprehensive assessment." Oxford Institute for Energy Studies (2009).

"Russia - International - Analysis." U.S. Energy Information Administration (EIA). Last modified July 28, 2015. https://www.eia.gov/beta/international/analysis.cfm?iso=RUS.

Schenk, Christopher J. "An Estimate of Undiscovered Conventional Oil and Gas Resources of the World." U.S Geological Survey Fact Sheet (2012).

"Uses of Natural Gas." Union of Concerned Scientists. Accessed March 12, 2016. http://www.ucsusa.org/clean_energy/our-energy-choices/coal-and-other-fossil-fuels/uses-of-natural-gas.html#.VuSKJZwrLIU.

Victor, Nadejda M. "Gazprom: Gas Giant Under Strain." Program on Energy and Sustainable Development (2008).

YOUR KNOWLEDGE HAS VALUE

- We will publish your bachelor's and master's thesis, essays and papers

- Your own eBook and book -
  sold worldwide in all relevant shops

- Earn money with each sale

Upload your text at www.GRIN.com
and publish for free